TOWN TO
COUNTRY

A GUIDE FOR TOWNSMEN WHO
SEEK A LIVING ON THE LAND

By

G. C. HESELTINE

CATHOLIC AUTHORS PRESS

Hartford, Connecticut

First published in 1933 by Burns, Oates & Washbourne, Ltd., London
Copyright ©2009 Catholic Authors Press, Hartford CT
Simultaneously published in Canada
www.CatholicAuthors.org
ISBN: 978-0-9783198-7-8

CONTENTS

BIBLIOGRAPHY

The following will be of interest to those who contemplate leaving the town for the country:

A New Policy for Agriculture, by F. N. Blundell. (Philip Allan, 7s. 6d.)

Horn, Hoof and Corn, by Viscount Lymington. (Faber & Faber, 6s.)

The Fairy Ring of Commerce, by Commander H. Shove, D.S.O., R.N. (Distributist League, 2S. 6d.)

The Agricultural Problem, by F. N. Blundell. (Sheed & Ward, IS.)

The Catholic Land Movement, Catholic Truth Society. *2d*.

Bulletins of the Ministry of Agriculture on various branches of poultry, pig, bee, vegetable, and fruit culture, at 6d. to IS. 6d. each, obtainable from H.M. Stationery Office, Kingsway, W.C.2, or from the Ministry of Agriculture, 10 Whitehall Place, S.W.I.

Land for the People, published Quarterly by C.L.A. of Great Britain, II4 West Campbell Street, Glasgow.

Liberty and Property, by H. H. Humphries. (Distributist League, IS. 6d.)

The Countryman, Quarterly, *2s. 6d*.

G. K.'s Weekly, Thursdays, 6d.

Cottage Economy, by William Cobbett. (Peter Davies Ltd.)

Farmer's Glory and *Hedge Trimmings*, both by A. G. Street. (Faber & Faber, 7S. 6d. each.)

Feeding Ourselves, by G. C. Heseltine. (*English Review*, April, 1933, 1*s*.)

The Farm and the Nation, by Sir E. John RusselL (Allen & Unwin, 1933, 7s. 6d.)

The Profitable Small Farm, by E. Graham. (Peter Davies, 1931, 7s. 6d.)

Dream Island (a record of the Simple Life), by R. M. Lockley. (Witherby, 8*s. 6d*.)

Britain's Trade and Agriculture, by Montague Fordham. (Allen & Unwin, 7s. 6d.)

Land and Life, by Lord Astor and Keith Murray. (1932.)

The Future of Farming, by C. S. Orwin. (Clarendon Press, 1930.)

TOWN TO COUNTRY

A GUIDE TO TOWNSMEN WHO
SEEK A LIVING ON THE LAND

I

FORWARD TO THE LAND

'The time has come, the Walrus said.'

E VEN politicians can now see, as a great many
normal and intelligent citizens have seen for
many years past, that it is necessary to
recultivate England. The loss of foreign markets
for our manufactures, the stagnation of the
manufacturing industries, the difficulty of
obtaining food from abroad in exchange for
manufactured goods that nobody wants, has
forced us to consider the possibility of growing
more of our own food. Now that the manufacturing industries can no longer employ the large
urban populations and no one expects that they
will ever do so again, we are considering the
possibilities of employing them in the
necessary work of growing food.

Cobbett said that 'it is abundant living among
the people at large that is the great test of good
government and the surest basis of national
prosperity and security.' The bad government
that is responsible for our present national
distress and insecurity has resulted in a
disturbance of that proper balance and
proportion between rural and urban life that is

essential to a reasonably stable civilisation, in which alone abundant living is possible. That bad government has been the natural consequence of the national state of mind. A nation of shopkeepers, business men whose sole business is to get rich quickly, has been too busy making money to attend to the business of living abundantly. Now that we can no longer make money we are beginning to turn our attention to more necessary and fundamental business.

So far as the Government is concerned, it is still obsessed with the business mentality. Plans for the restoration of agriculture are concerned with the secondary and rapidly fluctuating prices of foodstuffs rather than with foodstuffs themselves. Economists (whose science is uncertain and speculative, frequently contradictory) are regarded as the most important advisers on agricultural policy. At the present moment the chief Government measure for agricultural restoration is a Marketing Bill in which the business of buying and selling produce takes precedence over the more primary matter of producing it. So that if the system of marketing and distribution cannot cope with the foodstuffs produced, it is not that system which is to be improved, but production that is to be curtailed. This fantastic and topsy-turvy state of things is one of the consequences of listening to some nonsensical economic theory about over-production.

There can be no such thing as overproduction until every single individual in the country is fully assured of an abundant supply of necessary foodstuffs easily accessible. Until that happy state is achieved, what is wrong is not over-production, nor low price levels, but an improperly employed, ill-distributed and pauperised population.

Everybody now recognises that the time has come for something to be. done towards reviving agriculture and restoring a proper balance between town and country. In 1931 the Agricultural Land (Utilisation) Act was passed, which, though objectionable in many respects, provided for the settlement of unemployed persons on agricultural holdings, for training and equipping them to live on the land instead of exist on the dole. Bad economics again thwarted what good there was in that scheme, and the Act was suspended from operation on the false ground of financial stringency.

These matters are mentioned merely to show that the business of restoring British Agriculture has at last become urgent enough to claim government attention from what has hitherto been considered the far more important business of encouraging the manufacture of artificial silk stockings, and restricting the importation of American talkies. Further, there has been an increase of sound literature on agricultural matters, in which an intelligent rural idealism and proper appreciation of rural activity in relation to the national life has been supported by careful practical proposals for dealing with existing difficulties.

As is usual, well-informed opinion in the

country has been ahead of official activity. Those enthusiasts and idealists whose 'back-to-the-land' agitations during the past half-century got them labelled as cranks, have lived to see their propaganda widely accepted as common sense. There has always been a much larger body of opinion in favour of rural development, than has been generally recognised. Very many factory and clerical workers in the towns have yearned to live a quiet life, however hard worked, in the country, with some measure of independence. It has been a common thing for prosperous small tradesmen to retire to a cottage with a garden or a few acres away from the town. But the vast majority have lost the initiative or lacked the capital necessary for such a pioneering enterprise.

After the War of 1914-18, many ex-soldiers revolted against the idea of going back to the towns as button-pushers or clerks, and fought for their independence on smallholdings. A good deal of nonsense has been talked and written about smallholdings, on the one hand by theorists who were well enough off not to have to work on a holding, and on the other hand by Jeremiahs who had heard of failures, or those who lacked the requisite qualities and had failed themselves. Facts and figures were given on both sides, emphatically contradictory.

At the present time many men who are unemployed would rather work fifteen hours a day on their own smallholding, however precariously, than remain inactive on the pittance of the dole. The majority of the unemployed may be neither fitted nor willing to work a holding of their own, many of them may be unfitted or unwilling to be employed on the land. It is often assumed that the population of the towns would not live and work in the country because it has become too much attached to town life, and such 'luxuries' as cinemas. But the same racket has now reached the country-side. Some, indeed, may have become attached to the town slums and the town noise and the town dirt-to every mouth its lettuce,' as S. Jerome said–it remains highly probable that many are not so attached, and that many others would not take much weaning from the gaieties of Bootle or Bermondsey.

It is nonsense to regard a dirty and congested Social system, little more than a century old, as permanently essential to human happiness and for ever unchangeable. The system of herding people in large towns, forcing them to live on the smallest possible wage that they will accept without revolt, giving them the alternative of working for the anonymous masters of the cartels and combines, with intervals of doing nothing on the dole or starving to death, is a rotten system. It will be equally rotten when the same system is taken over by the state and called the Dictatorship of the Proletariat; in any case, the proletariat is never allowed to dictate. The employed man, under any system, does as he is told, and puts up with what he gets, kicking now and then to remind himself that he is a human being. It is nonsense to say that the mass of ordinary decent men and women, the bulk of the population of this country, will put up with such a system indefinitely for the sake of the movies and canned jazz. As the present economic condition of this country shows no signs of mending, or, what is more likely, it gets worse, we shall see an accelerating drift from the towns to the country, however unpromising the prospects of rural life.

As these words are being written, a Press report is to hand (for what such reports are worth) to the effect that in America (the other great victim of excessive urban concentration), 'it is estimated that more than 100,000 people have gone back to farms each year since 1930.' 'The farm population of Oklahoma has increased by 75,000 since the 1931 census.' 'Seventy per cent of the Texas unemployed have signified their desire to go "back to the land".' A most significant sentence in this Press item, an unconscious indication of the mind of the journalist, and typical of most men's minds on this matter, is as follows:'*With the general town-to-country exodus, a survey has shown that country folk are reverting to their former independence. Country" sociables" and schoolhouse "box suppers" are again the order of the farm family's life.*' Of course they order things queerly in America, but they are still human. And the fundamentals of this matter are common to human nature everywhere.

The drift from town to country in Britain is as yet slow but it is positive and increasing. So far it has no official encouragement. The destitute unemployed of the towns have no means of moving their homes and settling in the country, though some of them have tramped out into the country seeking odd jobs on the farms. But farm-workers themselves are unemployed owing to the depression of agriculture, so there is no room for new-comers. Now that there are some signs of an agricultural revival this condition of things may be mended a little in the future.

Meanwhile there are very many unemployed in the towns who are not included in the official figures of unemployed because they have not come within the scope of the unemployment insurance scheme. Their position is even more desperate than that of the workers in the grades below them because they cannot draw a dole. When they are unemployed they are completely destitute, once they have exhausted their savings or raised what little money they can on their small personal possessions. In this class suffering is acute. A certain sense of personal dignity and independence prevents them from applying for poor-law relief, which, indeed, they would not get until they had reduced themselves to utter destitution. It is amongst this class that there is a desire and willingness to work on the land, however hard and unaccustomed the conditions, rather than seek state relief, sponge on their friends, or starve with their families. Numbers of these stalwarts venture into the country-side whilst they still retain a little capital, feeling that they can at least grow a fair proportion of their own food. To this class must be added many of their fellows who have not yet lost their employment, but are daily expecting to do so, as a consequence of some new amalgamation or 'economy,' or the bankruptcy of their employers. Then again there are many members of the professional classes whose incomes are decreasing rapidly on account of the decrease in their clients' incomes–some of them are beginning to eat into their capital and savings and feel that it would be a sound plan to 'dig themselves in ' in the country, where they may have at least a fair chance of securing a reasonable amount of wholesome food from their own small bit of land, with a healthy life for themselves and their families.

It is in times of economic distress like the present that such men begin to put first things first and realise that food and shelter, self-respect and independence, are fundamentally more important that the fal-lals of town life.

'Fate cannot touch me: I have dined to-day.'

For such as these, this booklet is written to give them some idea of what they are in for, how to set about the change, pitfalls to avoid, and a few disillusionments. To be forewarned is forearmed. When personal fitness, experience and capital are none too plentiful, any tip that may prevent the waste of these valuable assets may mean all the difference between success and failure.

II

SMALLHOLDINGS

Happy the man Whose wish and care
A few paternal acres bound,
Content to breathe his native air
*   In his own ground.*
Whose herds with milk, whose fields with bread
Whose flocks supply him with attire;
Whose trees in summer yield him shade,
*   In winter, fire.*

ALEXANDER POPE.

Better boil herbs, thou toiler after gain!

THEOCRITUS, Idyll X.

BETTER boil anything, even your head, than toil after gain on a smallholding. Large holdings in Throgmorton Street are a more sensible proposition for toilers after gain, yet even they are no more bound to succeed than smallholdings at the other end of Watling Street are bound to fail. But when you have toiled after gain in vain, or you are no longer allowed to toil after gain, not even after somebody else's gain, you may think it better to boil herbs than to stew in your own juice; especially if they are your own herbs. You are far more likely to boil herbs successfully on a smallholding than in the City.

If, on the other hand, instead of toiling after gain you wish to do a little honest business, and trade hard work for a healthy independent life, a man's life, you will get your full money's worth on a smallholding. That is to say, if you work hard enough, and intelligently enough, you will get the abundant living which is your due. If your efforts are not worth much, you will not get much: you will fail.

There are those who will tell you that smallholdings are a failure. It is just as true to say that banks are failures, butchers' shops are failures, or publishing businesses are failures; except that smallholdings are never quite the failures that the other ventures are when they fail. For the smallholding does the smallholder some good honestly, even when he fails to make a financial success of it. And it does the country some good because more food has been produced on it than if the land were left idle, or worked in a more extensive and less concentrated manner by a general farmer. Smallholders sometimes fail for the same reasons that men fail at other ways of making a living. But on the whole the proportion of smallholders who fail is small. Smallholders nearly always fail to make fortunes, but few of them fail to make a living.

There are exceptions. Mr. Montague Fordham, in his very interesting book *Britain's Trade and Agriculture* (Geo. Allen & Unwin, 1932, 7s. 6d.), tells of a smallholder in the Fen country who, confessed to having made £15,000 in a very few years,' whether honestly or not we are not told, but you are not remotely likely to repeat that feat. That smallholder told of how, not very long ago, he had had twenty acres of potatoes, got ten tons per acre, and sold them for ten pounds a ton—he got £2000 for that one crop, and certainly pocketed £1000 profit, probably a good deal more. On the other hand, a neighbouring farmer, a year or two later, told of how he sold his last crop of potatoes for £1 a ton, and thought himself lucky to get it.

Mr. F. N. Blundell, himself a landowner and farmer, in his book A New Policy for Agriculture (Philip Allan & Co., 1931, 7s. 6d.), says: 'Smallholders are, in fact, making a living where large farmers are unable to do so (Report of the Work of the Land Division of the Ministry of Agriculture for 1924). The percentage of failures, where properly selected men have been placed on properly selected land, has not been high. The Report of a certain Smallholdings Committee to a County Council, for February 1931, mentions that arrears of rent since "the 1926 valuation" amounted to £177 12s. 10d. on a total rent roll of £13.489, or 1'31 per cent. Of how many private estates of comparable size could the same be said?

'There are districts in which the proportion of failures has been considerable and others in which it has been very small. But, taking all the schemes together, it would probably be true to say that the failures are much less numerous than in any of the post-war settlement schemes overseas. If the failures were analysed, it would be found that the cause lay, not in the size of the holdings, but in the unsuitability of the man or of the land. The pages of the County Councils' Gazette in 1929 and 1930, in which accounts are given of the experience of different counties, bear witness to the truth of what has been written above. They also supply much evidence that townsmen, without previous experience, are making good.'

Beware, by the way, of taking too much notice of figures and statistics on the subject of smallholdings (even such as these referred to by Mr. Blundell and indicated in the County Councils' Gazette). Figures are often misleading, firstly because of the recent wide fluctuations in values, secondly because they are never the whole story, thirdly because in the hands of the expert they can be used to prove anything. We will return to the point again when we come to consider the capital required and domestic economy.

The important thing about Mr. Blundell's view is that he considers smallholdings on the whole successful, and he emphasises the vital matter of the right man for the job.

In this his opinion is confirmed by general experience. Viscount Lymington, in his highly readable and interesting book *Horn, Hoof & Corn* (Faber and Faber, 1932, 6s.), says: 'The preliminary requisites for success *lie in having the right men and their wives for the work and then in having, first of all, the right type of agriculture for smallholdings* ....' And again: 'Smallholdings must depend then, not on theory or prejudice, but on the opportunity for the right. man to be a smallholder, and on the right place being chosen for him in which to own his smallholding.' Observe the significant word 'own,' to which we shall return later.

One more word on this point from Mr. Blundell:

'On the credit side of the smallholdings there is much to be said. New capital values are created in the country, and new rateable values. In the case of one smallholdings scheme it was estimated that the increased rateable value was equal to the proportion of loss, borne by the county concerned. The smallholder, even more than the large farmer, is a customer for the home manufacturer. He does not want the immense machines manufactured in the U.S.A. or Canada; he wants the smaller plant manufactured at home. He wants the implements he can see and buy in local shops.'

Smallholdings have been variously defined. A hundred acres begins to be a small farm. Most holdings are not more than fifty acres, a very large number are under ten. A man has been known to make a very good living on a holding of one acre, but you are not likely to do it.

The yield per acre on smallholdings is much higher than that on general farms, and the labour expended per acre must be higher also. Conditions vary so widely on smallholdings that statistics are generally misleading. The size of your holding will depend on the capital you have for buying it, the locality and quality of the land, and what you propose to do on it.

Rough pasture and waste land that is of very little value to the general farmer may be used to good advantage by the smallholder for pigs, poultry and goats, for example. Such land can be bought or rented cheaply.

Hitherto, smallholders have been discouraged and handicapped by a depressed and declining state of agriculture in general. If they have succeeded against such odds, the prospects of smallholders under an agricultural revival are good.

III

FARMS

Lord, 'tif thy plenty-dropping hand
That soils my land,
And giv'st me for my bushel sown
Twice ten for one,'
Thou mak'st my teeming hen to lay
Her egg each day,
Besides my healthful ewes to bear
Me twins each year,
The while the conduits of my kine
Run cream for wine.

ROBERT HERRICK.

IF you can afford it, you will go one better than
a smallholding and buy a farm. The
smallholding is essentially worked by the owner
himself, with the help of his wife and children.
The farm should also be a family affair, but it will
need extra labour unless the family is large and
farm-bred, in which case the farmer will not be
reading this.

The new farmer will not buy a farm until he
has had fairly comprehensive training,
preferably on a farm. He will not, if he is wise,
imagine that he can learn farming except by
farming. He can get a smattering of general
principles, and start with some useful notions
(if he keeps clear of the new pseudo-scientific
fads that will be pressed on him at the
agricultural colleges). But farming is a most
highly skilled profession, probably *the* most

highly skilled. It needs many generations to make a good farmer. If you are very keen and work hard, and have some measure of natural genius, you may make a tolerably good farmer after some years. But in the meantime you will need to rely on a good bailiff and good farm hands, bred to the life. The best farming demands not only a knowledge of farming processes, but a knowledge of the peculiarities of local conditions of the land and the climate. Experience alone gives that sort of knowledge. If you ask the locals they cannot tell you–not because they are ignorant, but because that sort of knowledge becomes instinctive with them: it never occurs to them to tell you what they take for granted and act upon without thinking about–they never imagine that it is of so much importance or that you are so ignorant as not to know it.

The general principles apply equally to a farm and to a smallholding, with the difference that on a smallholding the range of operations will not be so wide or on so large a scale. In spite of the theories of some simple-life enthusiasts who start with the idea that they should make themselves entirely self supporting, it will be found that with economic conditions as adverse as they are to-day, some measure of speciali-sation, if not absolutely essential, is certainly wise, both on the holding and the farm.

Single young men for whom neither the smallholding nor the general farm is within reach or practicable, will band together (as they have done already) for the rural adventure, until they can start on their own. For such companies a general farm, which allows for wide experience and distribution of labour, is the proper place. They, too, will get a good bailiff and a farm labourer or two to set the pace.

IV

ALLOTMENTS

A time there was, ere England's griefs began,
When every rood of ground maintained its man;

OLIVER GOLDSMITH.

D ON'T believe it. A poet who talks like that is taking liberties with his licence and ought to have it suspended. But if every rood of ground did: not maintain its man, many a rood had a good try. We all know very well, at least those greybeards among us who are old enough to remember the late War, that the roods of ground that were cultivated by the rude forefathers of the hamlet whilst we lived in the lap of luxury (in the person of the Army Service Corps) yielded a most valuable supply of food. Allotments are a very useful half-way house for the townsman who hopes to keep himself some day on land of his own. Let Lord Lymington, a practical farmer, speak again: 'The allotment is the one general place where the townsman in leisure hours can take part in life as an actor, rather than a spectator. The allotment is not only the provider of rations, but of new interests and creative self-respect. It can become the beginning of successful smallholding without the danger of painful and costly failure.'

Do not imagine that you can get all your experience for a smallholding on an allotment or in your vegetable garden. You can learn a lot and accustom your muscles to a little honest toil You can learn to use the spade, fork, rake, hoes and cultivator. You can learn to use these tools at a respectful steady pace, and to have them of a size and weight to suit yourself. You will learn that you do not dig most ground by going hell-for-leather at it with the biggest spade you can find. All that is worth knowing when you have to live on your own holding.

But you will find that there is an appreciable difference between digging or hoeing a few square yards in the evening on ground that is well worked, and digging all day, double-trenching rough fallow land, when you have to make a field into a garden. That will take the fire out of your enthusiasm for the first week or two. But it is one of the most worthwhile jobs in the world.

Experience on an allotment is never wasted, though you may decide to specialise in pigs or poultry or pink-eyed rabbits on your holding. For reasons that will appear in due course, you will need your own kitchen-garden whatever you do on a holding.

If you cannot see your way to leaving the town entirely it is a good plan to work an allotment or a part of your garden for vegetables. Food produced on an allotment is a great asset to the country at large. It reduces imports and eliminates transport and marketing costs. If there could be a revival of allotment cultivation as in wartime (to which the present crisis is comparable economically) it would make a decided difference to our situation.

Mention must be made here of the admirable work done by the Society of Friends' Allotment

Committee. 'The Society of Friends began relief work in the South Wales coalfields in 1926 by providing seeds, etc. for allotments at very reduced prices. The work was so fruitful that it received support from the Lord Mayor's Fund in 1929-30 and in 1930-31 the late Government assimilated and developed the work hitherto done by the Society of Friends and set up a Central Allotments Committee. During that year the number of allotment holders under the scheme (now extended throughout the country) was 64,000, and the value of the average yield in vegetables per plot from £5 to £7. About £400,000 worth of fresh food was thus provided by the voluntary and willing labour of unemployed men at a total net cost of about £23,000, which was provided by the Treasury.' That amount was presumably augmented by contributions from the men themselves.

During the past season (1931-2) the Government had gone economy-mad and, through listening to crazy economists, concluded that it could not afford to spend £23,000 in order to make £400,000 on a scheme already well tested and proved sound. The Government was too poor to accept a profit of some 1500 per cent, not to mention the physical and moral profit accruing to 64,000 citizens!

Hence the Society of Friends continued the work and in 1931-32 63,000 men produced about £380,000 worth of food at a cost of £30,000, half of which was contributed by the public and half by the men themselves.

National economy as well as common sense and the need for more abundant and cheaper garden produce, demands an immediate extension of such a scheme everywhere possible.

V

THE FAMILY UNIT

Housekeeping and Husbandry if it be good:
Must love one another as cousins in blood.
The wife too must husband as well as the man
Or farewell thy husbandry, doe what thou can.

THOMAS TUSSER.

Little children are like arrows in the hands of
the giant, and blessed is the man that hath his
quiver full of them.

'THAT,' says Mr. Cobbett, 'is a beautiful figure to describe in forcible terms the support, the power which a father derives from being surrounded by a family.' Nowhere does a man need that support more than on a smallholding or small farm. If a man is to work his holding to the full and give proper attention to the many things that have to be done to time, according to the season, he needs a wife who runs the home properly for him, and who is interested in the work of the holding sufficiently to give a hand outside when several things have to be done at once. There is a job to be done on the vegetable patch before the rain comes, the broody hens need to be let off the eggs and fed, the goats are waiting to be milked. That is were the family comes in useful. A smallholder cannot afford to pay for labour. Extra labour must be available at all sorts of odd times. If this sounds like regarding a wife as merely cheap labour, it must be remembered that the wife and family are partners in the concern, and profit sharers.

That is why it is absolutely essential for a townsman who ventures on to a smallholding to have a wife who is at one with him in preferring the country life, and in being ready to tackle hard work and long hours when necessary. If the partners in this pioneering adventure do not agree, the chances of success are very slim.

As for the children, they begin very early to be useful at small jobs of carrying and fetching. They have not so much to be amused (as in a town nursery) as to be allowed to amuse themselves by helping. There is enough variety on a smallholding to keep them interested. By the time they go to school, they can be very useful after school hours, and by the time they leave school they can be as valuable on the holding as most hired men.

Moreover, the education children will get by helping their parents on a smallholding is a good deal more interesting, and more valuable, than the eurhythmics and compound interest formulae that they will be taught at school.

What applies in this respect to the smallholding, applies equally well to the farm. The family is the natural and proper unit and as it grows it supplies local craftsmen for local needs and can spare a few vigorous healthy men, sane with the sanity of the soil, to improve the civilisation and culture of the cities and the State. If you cannot start your smallholding on a family basis, at least with a wife, you will do better to share somebody else's farm or holding, or join with other single pioneers in a community.

VI

THE COMMUNITY FARM

Some were employed in clearing the land of brambles and weeds; others in muck-spreading; others in hoeing or sowing; no one ate his bread in idleness.

Life of Herluin, Abbot of Bec.

THAT was written of the pioneer monks who founded the Abbey of Bec. Colonisation by communities (and we are about to colonise England) has nearly always been inspired by some religious motive. Such communities have always been held together by common ideals, usually religious—for example, the Pilgrim Fathers. Communities of young men who are inspired by the ideal of restoring the normal balance between town and country life, which must begin by an exodus from the towns to the country, have started community farms or colonies, to give practical expression to their social theories. They have gone on to the land to show that it can be done, that it is a good thing to do, that many more like themselves would be the better for doing it; and of course, they have undertaken this pioneer work as a means of supporting themselves and earning their own bread rather than remain unemployed or badly employed in the towns.

Such communities have undertaken high adventure. Several of them appear to be making a success of their farms. Normally these colonies will consist of single men, as at Old Brown's Farm, Chartridge, Bucks; yet families and single men may settle together in a loosely formed community so that they can afford one another mutual support and protection against the many enemies to success that will surround them in the wilds of rural England. One such colony has settled on a derelict estate at Langenhoe in Essex in an attempt to make that wilderness a fruitful and populous garden. If that colony succeeds, as it deserves to do and should do with proper organisation, against the incredible odds and in spite of the inexperience with which, it started, it will provide a magnificent example and inspiration. What that colony does at Langenhoe should be possible on thousands of similar barren and derelict areas throughout the country.

But the community farm, as at Chartridge, should be regarded as an expedient for training and helping men to establish themselves on the land, not as a permanent or ideal form of farming. For success, such a colony needs careful organisation and strict discipline. The system has only been historically successful in the case of the monastic orders, where a very strict discipline and organisation was found necessary. The modern community farm will need these things all the more since it will lack the deeper spiritual motive of the monks, though it may have a deep and sincere common religious inspiration. The other sort of colony is also an expedient, no doubt necessary for the pioneers. But it is a seedling, specially planted in unpromising conditions, of that normal social life of the country-side that must obtain in the future as it has done in the past if a healthy social condition is to prevail in the nation as a whole.

VII

Hie vir, hie est!

THE personal factor is by far the most important in determining the success of any venture from town life to country life, whether on a smallholding, a farm, or in a colony. That is why it is so dangerous to advise anyone, or encourage anyone to estimate their prospects of success by reading statistics about smallholdings, or accounts of what other people have done. In every trade or profession it is true that one man will succeed where another man will fail, under similar external conditions. But it is probably far more generally true of this adventure into rural life. In the town most trades and professions, having been acquired by the usual course of training, have a certain local demand, and succeed modestly at least on a moderate amount of conscientious work and application to the job; they are exposed only to local and economic risks that are in the main shared by every other similar trade or profession in the town.

The new smallholder or farmer, on the other hand, is handicapped by a far wider ignorance—there are more things for him to know and not knowing one of them at a critical moment may mean a heavy set-back. Then he has the weather to contend with, not only because it affects his crops and may ruin months of hard work for him, but because he has to endure its vagaries himself, personally. He has to be out in the weather, whatever it is. And being town-bred he will get aches and pains for which he would normally lay up if he lived in town. He must resist the temptation, or break the habit, of nursing his physical weaknesses.

The pioneer on the land will get many rebuffs and set-backs, due to his inexperience, to the weather, and to what looks like sheer bad luck, but is more likely ignorance. He must go on long after he is beaten.

It is not true to say, as so many soft townies say, that life on a smallholding or farm is a life of unrelieved drudgery. The continental peasant is often referred to with pitying scorn as a man who has to labour like a slave on his land all hours of daylight, for a mere subsistence. He does not need any pity. He is far more certain of abundant living than any factory worker. He is the least affected by the vagaries of world economics, the manipulation of exchanges. He feeds the towns, the academic and cultural centres of his system, bodily and intellectually. War and desolation can roll over him, and he comes up smiling a harvest or two after.

The urban worker who sets out to establish himself as a peasant in England has to create for himself in some measure a life which is already normal and established in peasant countries. He has to do it in the face of the adverse economic and social conditions obtaining in England, an industrial country. He is starting a new way of life for himself, unlike the natural peasant who has grown up in it. Therefore he must expect to find the task pretty gruelling at first.

The life will be gruelling both physically and mentally. He has to wrestle with the new and difficult conditions as much with his head as with his hands. That is why it is not enough for a man to leave the town merely because he is out of work, or because he is dissatisfied with his town job, or because he thinks he would like to live in the country.

'It will not serve,' writes Lord Lymington, 'to put half a million workers on the land as family farmers with all the softness of our present industrial system in the background. With nothing more than a perfunctory training in a job which often needs an inherited as well as a life education, it would be to court disaster. For the small farmer has not the big farmer's advantage. But like him he must be accountant and business man, vet, scientist, empiricist of the soil and weather, and organizer of labour in constantly shifting conditions. Yet an untrained man may, with supervision, learn the care of stock and test his endurance for a job which knows no hours, but all weathers.'

The same words apply to the smallholder, except that the labour he must organise is his own and his family's. The moderately well educated and intelligent sort of town man, whom we have in mind as a prospective smallholder, will also teach himself a good deal and learn more rapidly than the average unskilled labourer or factory worker, though he will probably lack their powers of physical endurance.

The supposed drudgery of life on a smallholding is only drudgery in the mind of the town worker whose work is normally so monotonous and uninteresting that he imagines all work is the same, and that work of any kind is in itself an evil to be eliminated at all costs.

The man who is 'going to make a success of a smallholding must be a man who is keenly interested in the work he has to do, or most of it (for there will be some drudgery in any occupation). He must be a man who can enjoy his work and get some fun out of it, some satisfaction in each achievement, in every job done. He will not then have any great sense of burden or drudgery in his work. He will be happier working ten or twelve hours a day, than the man who sits at a desk in an office for eight, spends two standing in a train between Croydon and the City, and two listening to banalities from Hollywood or Broadcasting House, playing bridge, or navvying in his garden.

Naturally the new smallholder must have a positive preference for open air, country sights and sounds, country peace, and country beer, rather than the stuffiness of clubs, the music of trams and roadbreaking drills, the bustle of constantly going somewhere in a great hurry only to go back again, and the doctored swipes churned through the pump in a London ' boozer.'

In addition to these matters of personal taste, the new smallholder or farmer needs a certain amount of aptitude for working with his hands.

He ought to be able to use simple carpentering tools, and get a certain amount of fun out of doing odd jobs. You cannot always get a carpenter or plumber at once by merely telephoning in the country—nor if you are trying to run a holding on small capital can you afford to do so. Nor can you afford to buy everything you need, in the way of hen-coops, fencing, pig-troughs, and the like. A man who can make small wooden things for himself has an enormous advantage, for the cost is usually less than half what you have to pay for the same things ready made.

You need to be reasonably fit and strong. A smallholding is no place for a pioneer who is in poor health or physically handicapped–unless he has extra capital available to employ help and compensate for the deficiency in fitness. The new way of life will give a man better health–but the worst of the struggle comes at the beginning, as a rule. Many broken ex-soldiers made good on smallholdings after the War, but they were assisted in many ways and had small pensions to rely on.

Age is another personal factor of importance. You may feel fit enough at forty-five or fifty, but the muscles will not readily work in the new ways. They can be made to do so, no doubt, gradually. Age is no bar to tackling a smallholding. But it may be a handicap of vital importance where speed is required in hard work. You must discount your personal potentialities according to age, and if necessary make some provision for assistance.

VIII

EXPERIENCE AND TRAINING

East is East, and West is West
(Though this may seem irrelevant)
Any damned fool can milk a cow,
But you can't muck about with an elephant.

ANON.

MR. ANON has said a good many things in his time, but he has not often tied up a fallacy and a truth so well. There are a great many damned fools, not all of them in Whitehall Place, who if it came to milking a cow would starve to death first. And the man who sets out to make a new life for himself on the land will undoubtedly find that he has to tackle something a good deal larger than a cow, which he will be wise to approach with no less respect than would be inspired by an elephant.

There have been poor men who have gone into the country all innocent and ignorant of what was in store for them, dug themselves in, and stayed there and prospered. You are no more likely to repeat their feat than Napoleon's when he took an army across the Pyrenees. If you want a gamble it will be cheaper and safer to buy a ticket in the Irish Sweep. In short, you must regard some training and experience as essential before you take a holding or a farm. Agriculture is the most highly skilled of all industries.

Training is chiefly of value in so far as it consists of experience. Theoretical study, book-learning, may have its own interest and a certain value. But it is worth nothing to the prospective countryman without being put "through the mill of practical experience. That must be obtained *before* you start on your own holding or farm, unless you can stand to lose much money and suffer bitter disappointments.

You can obtain training at various agricultural colleges throughout the country. The Royal Agricultural College at Cirencester; the Harper-Adams Agricultural College at Newport, Salop; the SouthEastern Agricultural College at Wye, Kent, and others provide comprehensive courses of training, of varying lengths, in various branches of agriculture on university lines at fees ranging from £120 p.a. inclusive to £150 p.a. with extras, for tuition, board and lodging, three terms of twelve weeks to the year, normal courses being three years. Inevitably these courses are largely academic for the training of teachers, research workers, and 'experts.' This is not to say that they do not include a good deal of practical work. At all these institutions actual farm work of all kinds is in progress and it forms part of the curriculum for students.

But by the nature of the case it is not farm work under the active service conditions of a real working farm or smallholding, any more than maneuvres on Salisbury Plain are the same thing as real war, however closely it may be attempted to simulate war conditions. The guns are loaded with blank ammunition and they are not fired by men fighting for their lives.

The same applies, though to a lesser extent, to the County Farm Institutes which many County Councils have established (you can get full particulars by writing to your local County Council). These farm institutes, however, are designed for the education and training of local agriculturists. The instructors are in the main very keen and competent persons, in touch with farmers throughout their own area, acquainted with local conditions, and in a position to advise accordingly.

The courses provided at these institutes are on the whole shorter, and they are cheap-fees ranging from £1 to £2 per week for full board, lodging, and tuition. They are by no means enough in themselves to make a town ignoramus a wise countryman, but they are very useful. Moreover, the instructors at these institutes are available to advise you and help you in your difficulties once you have started on your own. That is part of the official work of the Ministry of Agriculture, through the County Councils, for the development of the industry, and you should make full use of this assistance, for which you pay as a tax-payer.

Whatever sort of training of the academic kind you have, you should turn it into real experience by working on a farm. Working on a farm does not mean walking round with a farmer and watching other people work, but actually taking a job and working full hours at the hard work and dirty work of the farm. It will break you in physically (if it does not break you up), and it will be enormously instructive. You will learn things in the course of muck-spreading, plucking boilers against time, catching a colt in a six-acre field, or setting weaseland rat-traps, that you will not find in books.

There are difficulties in the way of getting your experience by working on a farm. The chief one is the farmer's refusal to have you on the farm at all. He cannot afford to have people blundering about and getting in the way of the men who are working. You may get a casual job at harvest or haymaking time and if you work hard enough the farmer may be persuaded to let you stay, though he is unlikely to pay you anything. Moreover, you must not do anyone else out of a job, or you will be made as uncomfortable as you deserve. If you know a farmer, or you can get a friend to recommend you to a farmer, so much the better. But if he does not make you work, you will both be losers.

Some farmers take pupils at a premium, or they may have you on the farm as a paying guest. This arrangement may be very good—it depends on the farmer. Beware of the farm that lives on its pupils, the farm that is a failure and finds pupils easier money than farming. There are plenty such farms about, especially poultry-farms and pigfarms, which might seem to be specially useful to the prospective smallholder. It is on account of these farms (where you mayor may not work, but you will learn nothing of any great value), that you should hesitate before going as a paying pupil on a farm, and you should not go to any farm merely because you have seen an advertisement. Otherwise you will waste time and money and make a fool of yourself in the end.

Wherever you go for training, and whatever you read, beware of too much' science' and too great an interest in the latest gadgets. By no means despise recent research, but make full use of the latest knowledge. But be sure it is knowledge. In the nature of things the more academic establishments tend to place too much stress on alleged new discoveries and labour-saving devices. New theories tend to be treated as discoveries or facts before they have been put to the hard tests of practical application over a long enough period of time. Many new theories are extremely plausible and they are supported by arguments that you, in your ignorance, cannot refute. It is when you try to put the theory into practice, on your stock or your crops, that you find its weakness—and by then it has cost you time and money.

Do not be bluffed by the impressive reports of exhaustive tests at the experimental farms. Scientists have weaknesses like any other men and one of them is for getting the result they want to get, especially if they are ambitious for notoriety. The best men are proof against that, but then you cannot tell the sheep from the goats yet. Of course it will be condemned as a foul and wicked slander to say that some scientists are not above recommending certain expensive methods of feeding as the most economical in the long run, and taking fees from the foodstuff maker as expert advisers. But it is true nevertheless. Be on your guard against these things.

Many mechanical appliances are heavily boosted whilst they are still in the experimental stage. As a beginner you cannot afford to subsidise the manufacturers in the experiments, by buying their gadgets and finding out their defects and then scrapping them and buying the improved pattern.

It is lamentable to have to suggest that you should do so little for progress as to wait until you have seen somebody else using a method or a gadget successfully for a good long time before you adopt it. But if you are to husband your resources you must let the richer men buy the experience and profit by their mistakes. A most reprehensible policy, no doubt, but a wise one.

Buying your experience by learning from your own mistakes and losses (mistakes always mean losses on a farm) is a very costly and disheartening business. You will be well advised to do as little of it as possible. You cannot avoid it altogether.

IX

CAPITAL

o Das Kapital!

KARL MARX.

O NE of the first questions that a prospective smallholder or new farmer asks is 'How much capital do I need to make a start?' It is a practical question of importance and before he starts he must have some sort of answer, however approximate. But no one can give him the answer except himself. There are people who will have the answer ready, with full details of facts and figures. But you may be certain it will be the wrong answer.

Since the great adventure of pioneering in rural England depends so much on the personal qualities of the pioneer, it is obvious that many personal factors will considerably modify the amount of capital required. Men have been known to go on to a holding, build themselves a rough hut, live on a diet that would not have shamed the hermits of the Thebaid, and work until they had established themselves as self-supporting 'men of property.' Not everybody can do that sort of thing, however much they would like to do it. Others have started with a few hundred pounds and succeeded in establishing themselves by dint of hard work and close living and a little luck with the weather or the market. The exact figures, or even approximate figures, are of no use to you.

The amount required varies widely according

to your experience, the type of holding you propose to run, its location, and the standard at which you and your family require to live. This last depends to a large extent on the standard at which you have been living. It is true that you can live at a much lower standard than that to which you have been accustomed if you are forced to do so by necessity; you can also reduce your standard in order to achieve your object of making yourself an independent countryman. But when this reduction is voluntary, it is not always as easy as it sounds. Moreover, it demands the co-operation of the whole family.

Until you have gained enough experience of the prices of land, and stock, and worked out your own rate of living, so that you can estimate for yourself what capital you will require, you are not fit to start on your adventure.

You will no doubt be told by various authorities that it is hopeless to think of starting with less than £500, £1000 or more. It is hopeless to start even with that if you have to take their word for it.

Something will depend, of course, on whether you buy your holding outright, rent it, or purchase it through a building society or through the local authority under the Small Holdings Acts. Your object will be to own your holding as soon as possible. But if your capital is very small, it is better to start on the other basis than not at all.

If you are handyman enough to build your own equipment such as poultry houses, pigsties, glasshouses and frames, you will need very much less capital than if you have to buy all these things ready made. A midway plan is to employ a local carpenter to work for you, or better still with you. But do not forget that whilst you are building your Own equipment in this way, you must add to the cost that of your own keep. And, of course, you will not be able to accommodate your stock so soon. Nevertheless, though it means a slower start, it is worth while if you are short of capital.

You can never buy ready-made poultry-houses, for example, as cheaply as you can make them. For one thing, when you make them you are saving transport costs and retailers' or wholesalers' profits. Theoretically a large firm of manufacturers should be still able to sell cheaper than you can make, because of mass production and machine manufacture. In practice they do not do so. Even if you do not save a lot of money by building yourself, you will be able to put in better material for the same money.

More technical apparatus, like incubators, is better bought than made if time is an important factor in your enterprise. You can avoid that sort of capital outlay, for example, by using broody hens—but you will be well advised to avoid that slow and laborious method, if you can afford to do so, because it will handicap your chances. This is not the place to go into the details of the how and why, but it is so. Certain schools of back-to-the-land primitivists strongly oppose making use of modern mechanical aids, on the ground that they are produced by, or inseparable from, the evil condition of our highly mechanised industrial civilisation—a condition which they are seeking to avoid and destroy when they go rural. For them, no doubt, that policy may be right—though it has all the appearance of going to extremes, a sort of fanatical hatred of machinery not only for the evils that are associated with it, but as evil in itself, which is but a recrudescence of ancient Manicheism. That contempt for, and rejection of, :machinery if purified of its Manichean tendencies may very well be part of the policy of those whose going rural is part of a semi-religious social crusade.

It is not, however, a practical policy for the man of limited capital, limited experience, and limited personal fitness, who seeks to create an independent livelihood on the land for himself and his family, for his own and his country's good. Since he is the man for whom this is written, he is advised to make full use of such modern mechanical aids to establishing his farm or holding on a sound basis, as he finds useful and can afford—having first, of course, made quite certain that they are useful and that they do not cost more than they are worth. It would be an impertinence, as well as foolishness, to attempt to advise crusaders for a cause. If that cause deserves victory, nothing that is written here can help or hinder it.

X

BUYING THE HOLDING—SITUATION

Mine be a cot beside the hill;
A bee-hive's hum shall soothe my ear;
A willowy brook that turns the mill,
With many a fall shall linger near;

S. ROGERS.

I will hold my house in the high wood
Within a walk of the sea ...

H. BELLOC.

THAT is all very well for the poets. You whose ambition it is to feed the poets (after you have fed yourself and your family) will have less choice in the matter. It is possible to choose your smallholding or your farm as a poet would, but it may mean that it will cost you a good deal in extra transport, time, and inconvenience. If your capital is limited, you cannot afford that luxury.

Do not flatter yourself that you can put up with a little extra inaccessibility for the sake of the poetically ideal situation. When that extra mile or unmade lane is constantly traversed in all seasons and weathers, it may prove the last straw, once you have got the hump, to break your back.

Moreover, the poet may be able to carry on with rushlights, water pumped up laboriously from a well, a bad postal service, and complete isolation. If you are to get your work done well enough and quickly enough to make your holding or farm self-supporting, if you need to get your surplus produce to market or to customers quickly, if you are a sociable fellow who likes company, you may have to put up with a modern horror of a dwelling-house, replete with modern conveniences, situated opposite a gas works.

Do not forget that every extra difficulty of access, every handicap of convenience, not only costs money but it is constant in its incidence. You cannot afford, even if you have a fair amount of capital, to add to the difficulties of a task that already bristles with difficulties that cannot be avoided.

As to whether you can buy a holding in working order, or whether you buy a piece of land and make a holding out of it, will depend on your own circumstances, chiefly on your capital and your choice of location. If you propose to market your produce direct, you must be reasonably near your customers. The country markets in their present condition are no outlet for the progressive smallholder or small farmer, except for products that are not good enough to market elsewhere. They are mostly worked by rings of middlemen dealing in the various products, and prices are apt to fluctuate, owing to local manipulation, in a manner disastrous to the small man. Officially organised marketing will be discussed later.

Meanwhile you will not concede too much in the way of practical conveniences of situation for aesthetic considerations. The temptation is great, but you cannot afford it.

Having decided where you want to live, so far as district is concerned, it is a good plan to visit the district, if you can, and see for yourself what land or property is available. Consult the local newspapers for properties advertised, and consult the local agents. But do not believe all, do not believe more than a quarter at most, of what they tell you about the properties they have to sell. They are born enthusiasts, poets if you like, prone to see beauty where ordinary folk like yourself can see none, and prone to rhapsodise about it extravagantly.

ALL THAT MOST DESIRABLE COUNTRY RESIDENCE SITUATE ON HIGH GROUND COMMANDING MAGNIFICENT VIEWS, FACING SOUTH, will almost invariably turn out to be a jerry-built corrugated-iron bungalow, or cheap brick and stucco villa, semi-detached, on a slag-heap, facing the gas-works. It is evens on its facing not south, but south-west, or dead in the face of the prevailing wind and rain. You may say that south should mean south and not south-west, but estate agents have a language of their own. Their mile, for instance, bears no relation to our statute mile of 1760 yards. If the agents' schedule says 'Station, post office, church, 2 miles,' those places may be anything up to 4 miles, but not less than 2½ miles away, even as the crow flies, and not as the bird hops or rides a bicycle or drives a cart or van, which is your immediate concern.

Before you waste time and money visiting properties recommended by estate agents, exclude everything they offer if it does not read exactly like the place you want. You will find even then that in ninety-five cases out of a hundred it is nothing like what you want. This is not an exaggeration or a libel on estate agents. It is the sober truth and you will save money by taking the tip.

Alternatively to purchasingΣ your own farm or holding, outright or through a building society or on mortgage, you may be able, if you are qualified according to the requirements of the Small Holdings Acts (see below, page 64), obtain your holding through the local authority. In that case you will obtain it on far more advantageous terms, financially, than you can obtain it in any other way, but you may not have so much choice of location or size of house or holding—conditions vary in different districts;' but if most of your capital is required for stocking and developing the holding, it will be worth your while to enquire about a holding from the local authority, i.e. the County Land Agent of the County Council, or the Town Clerk to the County Borough.

Your ultimate object, in any case, should be complete possession and ownership of your property as soon as possible.

XI

YE OLDE TUDOR COUNTREE COTTAGE

Low is my porch as is my fate
Both void of state ...
Like as my parlour, so my hall
And kitchen's small ;
A little buttery and therein
A little bin ....

ROBERT HERRICK.

AND that is how you come to crack your head on the lintel, you find your wife constantly complaining because she cannot indulge her favourite sport of 'swinging a cat round' in the kitchen, and you have to build extra storage because there is no room for anything in the buttery or the bin.

There is doubtless a positive advantage in living in a cottage that is picturesque and quaint, but that is not why you want the cottage. It may be an attraction for your friends in town—an asset of varying value. As a rule it will be snug, cosy and warm in winter, cool in summer. Our ancestors knew a trick or two that make for comfort. If it is in good condition, the cottage will be substantial, however lop-sided it may look.

But it will probably be expensive. It will cost more for the accommodation it gives than a new house built to your own plan; it will cost a good deal in unkeep—although it may be substantial enough in structure, the plastering is apt to need frequent attention, the roof of beautiful mossy old tiles is apt to fail in the purpose of a roof and let the rain in; you may find that rats and mice have a centuries-old game of hide and seek in the hollows of the walls or in the roof, difficult to get at; unless the cottage has been thoroughly 'conditioned' by competent workmen, you will find trouble with doors and window-frames. A cottage that has stood three or four centuries of English weather is bound to show signs of age and wear.

Not only that, many old cottages are apt to be damp, through having no damp-course. That may not matter to the ancient native stock that is bred in such cottages, but it will find the weak spots in, foreigners.' The cottage may lack sanitation, bath, or proper water supply. You may decide you can put that right. But deciding and doing are often far apart once you start work on the holding and you may find yourself living in constant inconvenience from those deficiencies. It is not so easy as it sounds for a townsman to put up with these things, breaking the habits of a lifetime. Moreover, you may find, when you start to remedy those defects, that you are hindered by by-laws, the proximity of other property, and so forth. If you decide to live 'primitive' you will find that it adds considerably to the normal difficulties of the new life—and some of the primitive tasks can be very irksome, not to say disgusting, despite your firm determination to go native.

If you still want Ye Olde Cottage, by all means have it, but rid yourself of your illusions about it first—and increase your chances of survival.

XII

DOMESTIC ECONOMY

Spouse of penniless Ibycus!

HORACE, OD. II. 15.

YOU have little capital, you are poor, you will go and live the life of a poor but happy countryman. Very well. The farm labourer, you argue, gets twenty-eight shillings a week and rears a family on it. If necessary you will live as he lives to gain your object and escape the town for ever. There, with your excellent resolution, you fall into grave error. It is as if you were to watch Mr. Maskelyne perform one of his wonderful conjuring tricks and resolve to go straight up on to the stage and do likewise. The miracle of domestic economy that is performed in the farm labourer's cottage has required a lifetime of training in an inherited craft, not to mention a fair amount of preparation behind the scenes—business with rabbits, for example, no less dark and mysterious than Mr. Maskelyne's.

The farm labourer ekes out his wages with the produce from his vegetable patch, which is old and well worked and will yield more than yours. Sometimes he has perquisites from the farm on which he works, waste potatoes and vegetables, or skim milk for his pig, chitterlings and other odds and ends of a pig or beast when it is killed—though these things come his way far more rarely in these days when farms are run on 'business lines' and the tendency is to treat the farm worker as a wage-slave on the same sort of basis as the factory slave. More's the pity. Making farms into factories, bringing the evils of the factory system, which is bankrupt, on to the farm, will do nobody any good. It all comes of listening too respectfully to the economists, professors of a most inexact and unreliable science.

Yet whether he gets them or not, farm perquisites do not exhaust the farm worker's resources, of which you know little and will never know much more. The finer art of rural scrounging has many wonderful secrets, which are not to be divulged here. But they are another reason why you will find it more difficult that you think to live on a farm labourer's basis, however noble and firm your resolve.

Furthermore, the local butcher, baker, grocer, blacksmith, plumber, carpenter, motor-engineer, will be, in their policy in dealing with you, of one mind, even if they are not one man, or one family, which has the same unanimous effect. That effect is that you will be charged more, if only a little more, for everything, than the farm worker will be charged. You, Ibycus, do not look like Mopsus, talk like him, walk like him, dress like him, nor in any way resemble him. Therefore you shall not be treated like him. In vain for you to say that you are penniless—no one is penniless in the country. Mopsus is not penniless; despite his poor wages he has a wad put away, and you who come from the town where the money is made, how can you be poor? Besides, it is only the gentry who talk about being poor, and hard-up. But they run up big bills, so a little extra is put on for the credit. Obviously you are 'one o' they.' So it, the little extra, shall be put on to you, too–if you do not want to run up a big bill, that is your affair.

And you, spouse of penniless Ibycus, they will stick the odd penny or twopence a pound on all your groceries, they will never have the cheaper grades of groceries, the cheaper cuts of meat (equal to any other in food value if not in prestige) for you. They are reserved for the spouse of Mopsus, who has always had them.

Therefore you must be prepared to find your household budget rather higher than that of your native neighbours. When making your estimate, put it on the high side and earmark some provision for an occasional break away, even a week-end in town, though you are quite sure you will never want to see the infernal place again, but will be perfectly happy communing with your cabbages for ever.

And when you are planning out the use of your capital, set aside enough to secure you the *necessaries*, on however modest a basis, for the first two years. You may not need it, then so much the better. But do not be so silly as to expect a living from your holding as soon as you have it stocked or planted, and expect at least one calamity.

XIII

THE SOCIAL SWIRL

'The more we are together, the merrier we will be.'

FROTHBLOWERS' ANTHEM.

O Beata Solitudo!

RICHARD ROLLE.

YOU hate the social swirl, Ibycus, you are sick of the sound of revelry by night, and the reaction of the morning after on finding that you have but provided copy for another modest half-guinea for a journalese Lord Snufflefuss or Lady Irene Twitterwit. That is why you have gone rural. What you want is the quiet labour of the fields by day and the fellowship of books by night, and be damned to the social swirl. Splendid! But is Mrs. Ibycus of the same mind? Excellent! Then you may read the Ayenbite of Inwit, or the Sentences of Peter Lombard to one another till the crack of doom, and you may skip the rest of this.

The human, normal, sociable fellow who likes company and diversion may want to know what to expect when he goes rural. So far as the canned entertainment of the towns is concerned, those who say the town worker will never be content with the dullness of the village, the lack of cinemas and the town vices, are living in the remote past. Unfortunately these things have penetrated to the country and every small country town, and many a village is fully equipped to supply the ninepennyworth of the Garbo or the Marx Brothers twice nightly. And you will be lucky if the bus does not pass your door. It is true that you will have less time for such riotous living, but you will have but little more difficulty in getting it if you want it. Note, however, that your bus fares will cost you more.

You may look forward to pleasant evenings in the village pub, playing darts and shove-ha'penny, and listening to rustic wisdom, gossip and weather lore. You will find, in all probability, that when you walk into the bar a dead silence will fall upon the company, as if you were suspected of being a secret agent of the Ogpu. You will make a few casual, friendly remarks, and receive respectful answers. You will then perceive that there are several empty pint mugs within your range of vision, and you will, if you are the man I take you for, pay up. After a further spell of polite and general conversation, mostly on your side, you will, again if you are the man I take you for, exercise your economic discretion and tact, and withdraw. It may be easier in the other bar, where your beer will cost you a penny more (for that town vice has also penetrated into the country). All of which should not surprise you, for after all, you are not 'one o' they,' and Mopsus knows it. Mopsus also knows his place if you don't know yours, and he will see to it that he keeps his, though you may not care a hoot about yours. He wastes nothing.

You will find that this arrangement has its value, especially when you want extra labour to help you when you are rushed. If you have persisted too strongly in upsetting his scheme of things, you will realise sooner or later that you have made a mistake. Don't run away with any silly ideas about snobbery. You will doubtless find Mopsus a snob at times. But what looks like snobbery is often only the preservation of status, one's personal position in the scheme of things that gives one the individuality and independence which is destroyed in the industrial towns, and which you have gone rural to recover.

You will find something more like snobbery on the other side of the village, unless you like to footle your time away playing bridge or golf and living above your means in order to destroy any impression that you need to do an honest day's work for your living. But that is not what you went rural for. So you will find that you are not 'one o' they' either.

Thus you will find yourself between what Mr. Bernard Shaw calls the 'skilly and carbidis.' If Mrs. Ibycus is as indifferent to that situation as yourself, well and good. After a few years you will have found kindred spirits and all will be well. But at first you will be neither fish, flesh, fowl, nor good red herring, and you will be treated accordingly. You will make the best of it; and in doing so you will find it useful to have a little provision for visiting the old cronies you have left in the Wen. You may tempt them to go rural, too.

XIV

SPECIALISATION

Jack of all trades, master of none.
Don't put all your eggs in one basket.

OLD SAWS.

THERE is a school of primitives, allied to those who have resolved to revert to the wooden plough because the modern plough is made of metal (which suggests machinery), who believe that when a man goes rural he should endeavour to be entirely self-supporting and wholly independent of any other human being in the world. In their view, the smallholder should strive to grow his own corn, his own meat, milk, eggs, wool, and leather and not only feed himself but makes his own clothes and boots. No doubt it would be good fun, and as a freak achievement of great difficulty, a matter of great selfsatisfaction. But such a policy is not normal, it has never obtained in any rural civilisation, and it is not likely to do so. It runs contrary to the normal social nature of man. It involves the fallacy that men cannot co-operate or be interdependent and free, when the fact is that only free men can properly co-operate and be decently interde-pendent.

When you go rural, if your aim is to make for yourself and your family a free, manly, self-supporting mode of life, you will not complicate an already difficult task by frittering away your energies in excessive subdivision and diversity. However many new activities you may care to undertake, for your own interest, after you are securely established, you will find that the most practical policy to pursue for a start is to have one main line of food production which will provide you with the income you require for paying rates and taxes and buying the things you cannot produce yourself. This applies to a small farm as well as to a smallholding, though naturally on a farm, in order to get the fullest use out of the land your production will be more varied in consequence of the necessity for rotation and so forth. Some farms, it is true, are almost wholly dairyfarms, in pursuance of the policy of livestock farming to which farmers have been driven by the depression in the corn markets; But corn growing must not be abandoned altogether. It is not true to say that it is not practicable in this country. We can and must grow corn if we are to build up and stabilise the agriculture of the country.

But for the smallholder corn-growing is not a practical proposition. Small livestock is at present far more profitable and gives a fair chance of success. Various branches of small livestock-farming, like pig-farming and rabbit-farming, have had serious ups and downs in recent years, though they may be stabilised by a general revival of agriculture. The same applies to fruit-farming. Market gardening (specialising in tomatoes and other fruits under glass, and so on) and poultry-farming, owing to the general constant need for the products and their perishable nature, have suffered also, but not so disastrously.

Whatever you adopt as the mainstay of your holding or small farm, will not, and must not, prevent you from producing other things in a small way for your own use, and selling the surplus. For example, if you have a poultry-farm (table egg-production is the safest branch to begin with) you will still need your own kitchen garden, largely because you may often find it difficult to buy vegetables in a village, except at town prices from town tradesmen who now tour villages with their vans. Your neighbours usually grow their own vegetables (and only sufficient for themselves), selling their surplus when they have any to town tradesmen or higglers who collect it. You will find it cheaper to grow all your own vegetables. You can keep a stock or two of bees, a goat or two for milk, and perhaps a sow. Similarly if you have a pig-farm, you will keep a few pullets to supply your own table with eggs and an occasional fowl.

But if you begin to develop any of these smaller side-lines on a larger scale you are almost certain to find that the different branches suffer from divided attention; so many things that take so long to do, want doing to both the poultry and the pigs, or the bees and the goats and the greenhouse at the same time. If you are on your own, with only a little assistance from the family, you will find that you have not enough hands, and the jobs do not get properly done. For example, it is both disheartening and disastrous to begin by making a larger kitchen garden than you can cope with. It will yield less than a smaller one.

XV

THE BUSINESS END

Eggs.-Per 120; Eng., ord. pkg. gr. I. 8s. 6d.-8s. gd.
Dch., 181b. mxd., 8s. 3d.-8s. gd.

MARKET REPORT 5/4/33.

WHICH being interpreted means that the best English eggs offered in that market fetched the same price as the best Dutch, partly because the Dutch were larger eggs and partly because they were better marketed. To remedy this, egg packing stations were set up under the Marketing Board for the purpose of marketing English eggs cleanly, tested and graded comparably with the Dutch. In theory the scheme is a good one. In practice the small producer gets the worst of the bargain. He suffers deductions for cracked and defective and dirty eggs, excessively (so he says). And it is easy to see that a packing station can handle the large bulk produced by the larger producer more economically, and will be subject to the normal human deference to the big fellow, and hesitate to annoy him. The running of the packing stations costs money, and it still remains to be seen whether under more reasonable conditions in agriculture generally, such an arrangement may be regarded as successful. This type of marketing organisation has recently been pushed on apace in several other branches of the industry. The Agricultural Marketing Act of 1931, and the Bill to implement and extend it now before

Parliament, introduce a dangerous measure of compulsion into the organisation of marketing, presumably because the schemes are such that their only chance of operating successfully is by handling the whole of the produce in a given area, and killing anything in the nature of competition.

In addition to the grave objections to compulsion of any kind, only justifiable by most serious national emergency and the proved absence of sound alternatives, from our point of view of pioneering smallholders or small farmers the liabilities under the new Marketing Schemes are serious. In the event of failure or unsatisfactory working, the larger producers are in a stronger position to bring about a winding-up, and since debts and losses of the scheme will be borne in proportion to the business done through the scheme, or on some similar basis, it is easy to see that a loss of, say, £50 borne by a smallholder may be a much more serious matter to him than one of £500 borne by a man with ten times his turnover.

The marketing schemes may relieve the small farmer and smallholder of much concern about prices, because they will tend to fix the prices of raw materials as well as produce, and leave him no choice or need for decisions in the matter. But that, in any case, will probably only apply to his main product, or any product exceeding a certain amount. He may still have a good deal of business to do on his own account.

Where it is possible it is a very sound plan to market your surplus produce direct in the nearest town. It will give you a lot more work, but once you have shown your customers that your produce is fresh and of good and constant quality, you will be relieved of much of the uncertainty of any other form of marketing, and your margin of profit will be higher. It may be urged against this that it is not the producer's business to be a shopkeeper as well, but when the markets are in such a condition that being your own salesman makes the difference between success and failure on your holding or farm, your business is to do whatever legitimately secures the end you have in view.

A good deal of conflicting discussion has taken place about co-operation amongst smallholders and small farmers. Certain co-operative organisations, such as the egg societies of Preston, Framlingham, and the Golden Vale, have been successful. But on the whole, except where production of a certain kind is fairly concentrated, co-operation in agriculture has had many failures and proved very difficult. If you can get together with a few neighbours to buy raw materials in bulk on the cheap, so much the better. But you will find many obstacles—hence the desperate measures by the Government for compulsory co-operation under a new agricultural bureaucracy, a remedy that may prove worse than the disease.

In the matter of transport, it will almost certainly pay you to own your own and not depend on local carriers. Now that small cars and vans are cheap, the convenience they provide and the time and labour they save are usually well worth the cost. To the anti-machine crusaders this will sound like advocating living on and supporting the foul industrial system that you have turned your back upon. But your business is to build your castle and arm yourself, and if you find that using the weapons of the enemy enables you the more quickly and certainly to put yourself in the position of beating him, then use them.

APPENDIX

I

FOR detailed advice on the prospects and possibilities of obtaining land in any particular area, apply to the County Land Agent or the Secretary of the County Agricultural Committee for the area, addressed through the offices of the County Council. From the same source you may also obtain particulars of training courses at the nearest County Farm Institute, where the instructors will be available to advise you.

The country is divided into Provinces (for the purpose of Agricultural Education) and advisory officers are attached to each provincial centre. A list is supplied at the end of Mr. F. N. Blundell's booklet The Agricultural Problem, but you can no doubt get any information you require on this point by applying as advised above.

2

The Catholic Land Associations are established in various parts of the country. They are doing most important and valuable work and you might find them useful.

The local secretaries are as follows :

South of England Catholic Land Association:
    Hon. Sec., Bryan Keating, 37 Norfolk Street, Strand, W.C.2, or Old Brown's Farm, Chartridge, Bucks.

North of England Catholic Land Associations:
 Liverpool: Hon. Sec., James Gavin, 134 Strand Road, Bootle.

 Salford: Hon. Sec., J. H. Cosgrove, 6 Mossgrove Road, Timpedey, Cheshire.

 Midlands: Hon. Sec., H. Robbins, Weeford Cottage, Sutton Coldfield.

Scottish Catholic Land Association:
 Hon. Sec., J. P. Magennis, II4 West Campbell Street, Glasgow, C.2.

 Enquiries will also be welcomed by Mr. John Hawkswell, Langenhoe Hall, nr. Colchester, where it is understood that land and work is available for volunteers.

3

 For unemployed and others without adequate means for going on to a holding of their own, the Society of Friends Allotments Committee should be approached through any local allotment society or through the Secretary, Mr. Fred E. Dodson, Friends House, Euston Road, London, N.W.I. 'Briefly, the scheme offers to every unemployed, partially employed, or seriously impoverished worker the requisite seed potatoes, vegetable seeds, lime, fertilisers, and tools. In order to receive this benefit the men are asked to group themselves into Allotment Societies affording opportunities for mutual helpfulness and co-operation.'

4

Circular of the Ministry of Agriculture and∑ Fisheries on the provision of smallholdings, etc., which may be had on application to the Secretary, Ministry of Agriculture, 10 Whitehall Place, S.W.1:-

MINISTRY OF AGRICULTURE AND FISHERIES

PROVISION OF SMALL HOLDINGS, COTTAGE HOLDINGS, AND ALLOTMENTS BY COUNTY COUNCILS, AND COUNCILS OF COUNTY BOROUGHS

Facilities are available whereby men and women with suitable qualifications (see below) may obtain small holdings and cottage holdings under the Small Holdings Acts.

Definitions.-A small holding for the purposes of those Acts means a holding above one acre and either not exceeding 50 acres, or, if exceeding 50 acres, then not exceeding an annual value of £100 for the purposes of income tax. It may be either bare land or equipped with buildings, including a dwelling house.

A cottage holding means a dwelling house, together with not less than 40 perches and not more than one acre of agricultural land which can be cultivated by the occupier of the house and his family.

Qualifications.-An applicant for a small holding must possess (1) sufficient experience to enable him to cultivate a holding of the type required; it is useless for an inexperienced man or woman to apply; (2) adequate capital to stock and cultivate the holding, no provision being made for grants or loans from public funds for these purposes.

A Council is empowered to provide a cottage holding for any person who is, in the opinion of the Council, a suitable person, and who satisfies the Council that he will reside permanently in the dwelling house comprised in the holding and that he has the intention, knowledge and capital to cultivate satisfactorily the land forming part of the cottage holding.

Where to apply.-Application for a small holding or a cottage holding should be made to the County Land Agent of the Council for the County in which the holding is desired. Generally speaking, councils will give preference to resident applicants, and, consequently, if application be made to the Council of some other county than the one in which the applicant resides it will be liable to be postponed.

Applicants will, as a rule, be interviewed before being accepted; they may then have to wait some time before a holding can be provided for them.

If the applicant resides in a county borough application should be made to the Town Clerk.

Tenure.-An applicant for a small holding, whose application is approved, may either:

(1) rent the holding from the Council either on annual tenancy or on lease for a definite term of years at its full fair rent for agricultural purposes: or, (2) purchase the holding, if the

Council own it, and are prepared to sell it. In that case the consideration will be the payment of (*a*) an annuity equal to the full fair rent of the holding for a period of 60 years, or (*b*) if the small holder desires, an annuity for a shorter period of equivalent capital value.

In the case of a tenancy, the tenant will be required to pay, on entry, for any tenant right or compensation for improvements payable to the outgoing tenant. He will also have to observe certain conditions for securing the proper use and good cultivation of the holding, and, subject to conformity with these conditions he will enjoy security of tenure. If a tenant quits his holding by his own wish, he will be entitled to recover from the Council compensation for improvements, unexhausted manures and feeding stuffs in accordance with statutory provisions.

Where the holding is purchased, the purchase annuity will be payable in half-yearly instalments, the first of which is due on completion of the purchase. A purchaser, equally with a tenant, will be required to pay, on entry, any compensation due to the outgoing tenant for tenant right and improvements. Unlike a tenant he will also have to pay any outgoings chargeable on the land and to bear the cost of all repairs. A purchaser must also fulfil the conditions which a tenant is required to observe for securing the proper use and good cultivation of the holding. When once the annuity is paid off, however, the purchaser, or his successor, will enjoy the full ownership of the holding. For a period of 40 years from the date of the sale, and thereafter as long as the annuity is undischarged, certain restrictions will apply to the use and disposal of the holding.

A purchaser may, by a bequest, leave his holding to some one person who will be responsible to the Council for discharging all the obligations of the purchaser.

The terms for renting or purchasing a cottage holding are similar to those relating to a small holding.

Advances for the Purchase of Existing Holdings. As an alternative to the purchase of a small holding provided by a Council, any person desirous of purchasing an existing small holding from a private owner, who is willing to sell him the holding, may apply to the Council for an advance, and the Council may, if they think fit, lend him an amount not exceeding nine-tenths of the value of the holding. The Council must be satisfied that the title to the holding is good, that the sale is made in good faith, and that the price is reasonable. Any advance so made is repayable by a terminable annuity for a period not exceeding 60 years. A holding purchased in this manner is held subject to the same conditions and restrictions as are applicable to a holding provided and sold by the Council.

An existing cottage holding may be purchased with the aid of a similar advance by any person who is qualified to be provided with a cottage holding (see p. 64).

Advances for the purpose of Equipment. – The purchaser of a small holding provided by, or purchased with the assistance of, a Council, may obtain from the Council, or from a Society under the guarantee of the Council, a loan for the purpose of enabling him to construct, alter or adapt a house or farm building on the holding. The Council must satisfy themselves that the case is in every way suitable, and, if so satisfied, may advance up to nine-tenths of the value of the borrower's interest in the property. Interim advances may be made as the work proceeds. Advances are repayable by instalments on terms arranged with the lenders.

Postponement of Payment of Instalments.- A council may postpone, for not more than five years, the payment of any part of the terminable annuity (except what is payable on completion) in consideration of capital expenditure by the purchaser which in the opinion of the Council increases the value of the holding.

Co-operative Tenure of Small Holdings.- Councils are empowered to sell or let one or more small holdings to persons working on an approved cooperative system or to an association formed for the purpose of creating or promoting the creation of small holdings, and so constituted that the division of profits amongst the members of the association is prohibited or restricted.

ALLOTMENTS

Any man or woman, who, by reason of having insufficient capital or experience or otherwise, is not qualified to take a small holding, or a cottage holding, but who is able and desires to cultivate a small piece of land in his or her spare time, may apply to the local allotments authority for an allotment. The allotment authority in a borough or urban district is the Council for the borough or urban district, and in a rural parish the parish council, or, if there is no parish council, the parish meeting.

Allotments provided by these authorities can only be rented. The rent payable is the full fair rent of the land for use as allotments.

5

The Agricultural Land (Utilisation) Act, 1931, provided for training and settlement on the land, on smallholdings and allotments, with financial assistance, of unemployed persons and agricultural workers and the establishment of training centres in connection therewith. This Act 'is in abeyance owing to the overriding necessity for economy in national expenditure, and the Ministry is unable, therefore, to take any action in the matter at present.'

Rather than spend money on establishing unemployed on smallholdings where they can produce food at a great profit (as the Society of Friends Allotment Scheme has shown), the Government thinks it more economical to spend that money in keeping the same men in idleness, producing nothing but despair and discontent. Every effort should be made, by wide expression of public opinion, by pressure on Parliament through members, by press propaganda, to end that extravagant economy and start the more fruitful economy, by operating Part II of the Agricultural Land (Utilisation) Act, 1931.

6

A new Private Member's Bill recently introduced into Parliament, Mrs. Ward's' Home and Empire Settlement Bill' seeks to extend the benefits of the Empire Settlement Act, 1922, *to persons who are desirous of engaging in agriculture in the United Kingdom*; and to remove any doubt as to whether it is in the power of the Secretary of State to contribute to the expenses of training and/or settling persons in the United Kingdom or Overseas under a Government scheme or any other approved scheme.

This bill provides for making available for these purposes sums not exceeding one and a half million pounds in the first year, or three million pounds in subsequent years, and it would appear that it might provide the necessary deserving support to such bodies as the Friends Allotments Committee and the Catholic Land Associations, and bodies with similar policies in view.

7

The Birmingham Branch of the Distributist League, Hon. Sec. K. L. Kenrick, 7 Soho Road, Handsworth, Birmingham, has prepared an interesting and careful scheme for settling unemployed workers on the land. It will well repay study. Copies may be had on application to the Hon. Sec. as above, or to the Hon. Sec.

APPENDIX

Distributist League, 2 Little Essex Street, London, W.C.2.

Catholic Authors Press

Catholic Authors Press is dedicated to promoting and preserving our rich Roman Catholic literary heritage. Catholic Authors does this through the rescue and recirculation of used and out-of-print books as well as the publishing of rare and classic titles from the past for the next generation of faithful. Catholic Authors maintains a comprehensive biographical database of Catholic writers accessible online and plans to offer workshops for young and aspiring Catholic writers in answer to the plea of Pope Pius XI: "In vain do you build schools and churches if at the same time you do not also build up a good Catholic literature."

www.ingramcontent.com/pod-product-compliance
Lightning Source LLC
Chambersburg PA
CBHW032028040426
42448CB00006B/759